NATIONAL GEOGRAPHIC KiDS

美国国家地理
双语阅读

Sea Turtles
海龟

懿海文化 编著

马鸣 译

第三级

外语教学与研究出版社
FOREIGN LANGUAGE TEACHING AND RESEARCH PRESS
北京 BEIJING

京权图字：01-2021-5130

图书在版编目 (CIP) 数据

海龟：英文、汉文／懿海文化编著；马鸣译. —— 北京：外语教学与研究出版社，
2021.11（2023.8 重印）
（美国国家地理双语阅读. 第三级）
书名原文：Sea Turtles
ISBN 978-7-5213-3147-9

Ⅰ. ①海… Ⅱ. ①懿… ②马… Ⅲ. ①海龟－少儿读物－英、汉 Ⅳ. ①Q959.6-49

中国版本图书馆 CIP 数据核字 (2021) 第 228172 号

出 版 人　王　芳
策划编辑　许海峰　刘秀玲　姚　璐
责任编辑　姚　璐
责任校对　华　蕾
装帧设计　许　岚
出版发行　外语教学与研究出版社
社　　址　北京市西三环北路 19 号（100089）
网　　址　https://www.fltrp.com
印　　刷　天津海顺印业包装有限公司
开　　本　650×980　1/16
印　　张　37.5
版　　次　2022 年 3 月第 1 版　2023 年 8 月第 4 次印刷
书　　号　ISBN 978-7-5213-3147-9
定　　价　188.00 元（全 15 册）

如有图书采购需求，图书内容或印刷装订等问题，侵权、盗版书籍等线索，请拨打以下电话或关注官方服务号：
客服电话：400 898 7008
官方服务号：微信搜索并关注公众号"外研社官方服务号"
外研社购书网址：https://fltrp.tmall.com

物料号：331470001

记载人类文明
沟通世界文化
www.fltrp.com

Table of Contents

A Sea Turtle!

Green sea turtle

Q What do you call a sea turtle that flies?

A A shell-icopter!

What hatches on land but spends its life in the sea?

What starts out the size of a Ping-Pong ball but can grow up to seven feet long?

A sea turtle!

5

Ocean World

Leatherback sea turtle

Sea turtles are graceful swimmers in the water. Their flippers move like wings.

Turtle Term

REPTILE: A cold-blooded animal that lays eggs and has a backbone and scaly skin

Sea turtles travel the world in warm ocean waters. They are one of the few reptiles that live in the sea.

A sleek body helps
the turtle move easily
through the water.

The scales
on its shell are
called scutes.

The back flippers steer
the turtle as it swims.
They are also used to dig
nests in the sand.

Green sea turtle

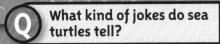

A sea turtle has lungs because it breathes air. A sea turtle holds its breath underwater.

Sea turtles can't pull their heads and limbs into their shells like land turtles can.

Their large, powerful flippers act like paddles.

Scientists believe some sea turtles live 80 years or more, but they don't know for sure.

Meet the Turtles!

The loggerhead is the most common sea turtle in the southeastern United States.

There are seven kinds of sea turtles in the world. Each has special features.

The flatback has a flat body. It lives near Australia.

The olive ridley has an olive-colored shell. It is shaped like a heart.

The hawksbill can't dive deep. It spends most of its time on the water's surface.

The green turtle has a small head. Unlike other sea turtles, it goes ashore to warm itself in the sun.

The Kemp's ridley likes shallow waters. It's the world's most endangered sea turtle.

Turtle Term

ENDANGERED: At risk of dying out

The leatherback doesn't have a hard shell. Its skin is rubbery with small bones underneath.

13

Nestbuilding

Female olive ridley sea turtles

A female sea turtle comes on land to lay her eggs. She usually returns to the same beach where she hatched.

Scientists aren't sure how sea turtles know where to go. They think sea turtles know by instinct.

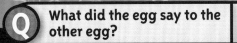
The sea turtle digs a hole with her back flippers. She lays her eggs and covers them with sand. Then she returns to the sea.

Turtle Term

INSTINCT: Behavior that animals are born knowing how to do

Female green sea turtle

15

Oh, Baby!

CRAAACK! The eggs hatch after 50 to 70 days. Tiny turtles called hatchlings crawl out of their eggshells.

Turtle Term

HATCHLING: A baby animal that has just come out of its egg

They are less than three inches long.

3 INCHES

Baby loggerhead sea turtle hatching from its shell

Hatchlings usually crawl toward the sea at night. In the dark, they are hidden from predators.

The little turtles follow the brightest light. The line where the sky meets the sea is the brightest natural light on a beach.

If the hatchlings follow this light, they will make it to the sea.

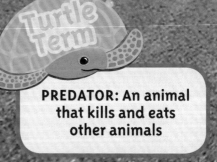

PREDATOR: An animal that kills and eats other animals

Q What do you get when you cross a turtle and a porcupine?

A A slowpoke!

Leatherback hatchling

Big and Small

The smallest sea turtles are the Kemp's ridley and the olive ridley. Adults are about two feet long and weigh up to 100 pounds.

Kemp's ridley sea turtle

Leatherback sea turtle

The largest sea turtle is the leatherback. It can grow up to seven feet long and weigh about 2,000 pounds.

On the Menu

Green sea turtle

Munch, munch, what's for lunch?

Most sea turtles eat plants and animals. They dine on algae (AL-jee) and sea grasses. They also munch on crab and conchs.

Turtle Term

ALGAE: A simple plant without stems or leaves that grow in or near water

Jellyfish are a favorite food for many sea turtles. But plastic trash can look like jellyfish in the ocean, and that spells trouble! Swallowing trash can hurt and even kill sea turtles.

Green sea turtle

Danger!

Hawksbill sea turtle caught in a net

Trash isn't the only danger to sea turtles. Fishing nets and hungry animals can harm them, too.

Building lights confuse hatchlings so they don't reach the sea.

Leatherback sea turtle confused by city lights

Sometimes people even step on sea turtle nests by accident.

Sea Turtle Rescue

In 2010 a giant oil spill leaked into the Gulf of Mexico. Oil covered sea animals and washed up on beaches. Oil is dangerous to people and wildlife.

Oil on beaches in Louisiana

Oil-covered Kemp's ridley

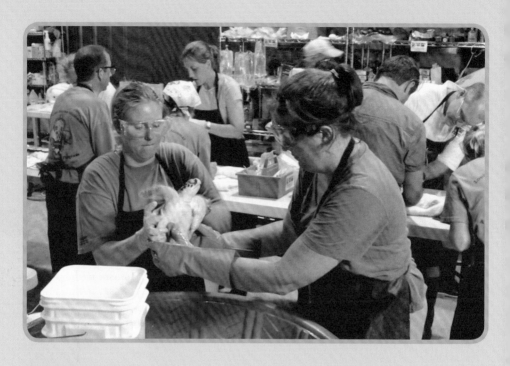

People in charge of a sea turtle rescue program in Louisiana saved many sea turtles.

The rescuers cleaned the turtles and gave them medicine. People cared for them until they could return to the sea.

Kemp's ridley sea turtle

Safekeeping

You don't need to work at a sea turtle hospital to help sea turtles. Here are a few things you can do to keep them safe.

1

Pick up trash on the beach.

2

Don't release balloons into the air. (They often end up in the sea.)

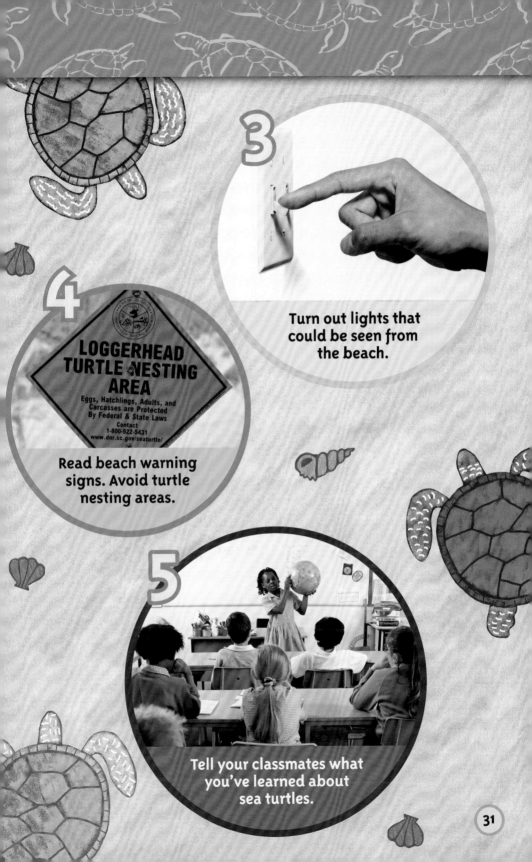

3

Turn out lights that could be seen from the beach.

4

LOGGERHEAD TURTLE NESTING AREA

Eggs, Hatchlings, Adults, and Carcasses are Protected By Federal & State Laws

Contact
1-800-922-5431
www.dnr.sc.gov/seaturtle/

Read beach warning signs. Avoid turtle nesting areas.

5

Tell your classmates what you've learned about sea turtles.

Glossary

ALGAE: A simple plant without stems or leaves that grow in or near water

ENDANGERED: At risk of dying out

HATCHLING: A baby animal that has just come out of its egg

INSTINCT: Behavior that animals are born knowing how to do

PREDATOR: An animal that kills and eats other animals

REPTILE: A cold-blooded animal that lays eggs and has a backbone and scaly skin

▶ 第4—5页

海龟！

绿龟

什么在陆地上孵化，却生活在海里？

什么出生时只有乒乓球大小，却可以长到 7 英尺（约 2.13 米）长？

海龟！

▶ 第6—7页

海洋世界

在水中，海龟是优雅的游泳者。它们的鳍状肢像翅膀一样活动。

海龟在温暖的海域环游世界。它们是为数不多的生活在海中的爬行动物之一。

棱皮龟

海龟小词典

爬行动物：长着脊柱和鳞状表皮的卵生冷血动物

▶ 第8—9页

流线型的身体有助于海龟在水中活动。

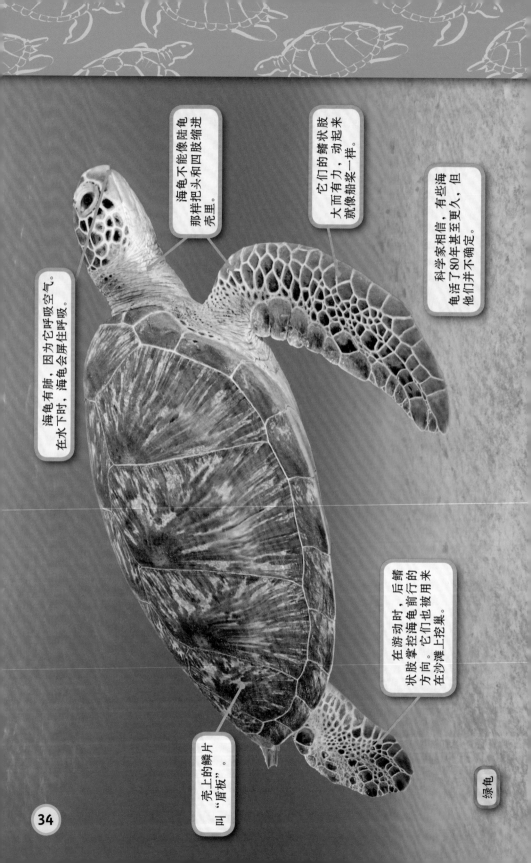

海龟有肺，因为它呼吸空气。在水下时，海龟会屏住呼吸。

海龟不能像陆龟那样把头和四肢缩进壳里。

它们的鳍状肢大而有力，动起来就像船桨一样。

科学家相信，有些海龟活了80年甚至更久，但他们并不确定。

在游动时，后鳍状肢掌控海龟前行的方向。它们也被用来在沙滩上挖巢。

壳上的鳞片叫"盾板"。

绿龟

34

第 10—11 页

认识一下海龟吧！

世界上有七种海龟，每一种都有自己的特点。

平背龟的身体是扁平的。它生活在澳大利亚附近。

蠵龟是美国东南部最常见的海龟。

太平洋里德氏龟的壳是橄榄色的。它的外形看起来像一颗心。

大西洋里德氏龟喜欢浅水区。它是世界上濒危系数最大的海龟。

第 12—13 页

玳瑁不能深潜。它大部分时间都生活在水面上。

海龟小词典

濒危：有灭绝的危险

绿龟的头很小。与其他海龟不同，它会爬到海岸上晒太阳取暖。

棱皮龟没有坚硬的壳。它的皮肤富有弹力，皮肤下有小骨头。

▶ 第 14—15 页

筑巢

　　雌性海龟来到陆地上产卵。她通常会回到自己孵化的海滩上产卵。

　　科学家不清楚海龟是怎么知道该往哪儿去的。他们认为海龟是靠本能知道的。

　　海龟用她的后鳍状肢挖洞。她产下卵，用沙子盖住它们。然后她回到海里。

海龟小词典

本能：动物天生就知道该怎么做的行为

雌性太平洋里德氏龟

雌性绿龟

▶ 第 16—17 页

哦，宝宝！

　　咔嚓！50 到 70 天后，卵孵化了。小海龟，也叫"幼体"，从卵壳里爬出来。

　　它们不到 3 英寸（约 7.62 厘米）长。

正破壳而出的蠵龟宝宝

海龟小词典

幼体：刚刚从卵里出来的、年幼的动物

海龟小词典

捕食者：杀死并吃掉其他动物的动物

▶ 第 18—19 页

　　幼龟通常在晚上爬向大海。在夜色中，它们可以躲过捕食者。

　　小海龟追随着最亮的光线。海滩上最亮的自然光就是海天相接的海平线。

　　如果幼龟跟着那道光爬，它们就可以成功地爬到海里。

小棱皮龟

▶ 第 20—21 页

大和小

　　最小的海龟是大西洋里德氏龟和太平洋里德氏龟。这两种成年海龟大约 2 英尺（约 60.96 厘米）长，体重可达 100 磅（约 45.36 千克）。

大西洋里德氏龟

棱皮龟

　　最大的海龟是棱皮龟。它可以长到 7 英尺（约 2.13 米）长，体重约 2,000 磅（约 907.18 千克）。

▶ 第 22—23 页

菜单

　　嚼呀，嚼呀，午餐吃什么呢？
　　大多数海龟吃植物和动物。它们以藻类植物和海草为食。它们也吃蟹类和贝类。
　　水母是许多海龟最喜欢的食物。但海洋里的塑料垃圾看起来很像水母。那就麻烦了！吞下垃圾会让海龟受伤，甚至死亡。

绿龟

海龟小词典

藻类植物：生长在水里或水边、没有茎或叶的简单植物

绿龟

▶ 第 24—25 页

危险！

垃圾不是海龟面临的唯一危险。渔网和饥饿的动物也伤害着它们。

建筑物的灯光会迷惑幼龟，使得它们无法爬到海里。

有时候，人们甚至会在无意中踩到海龟的巢。

被渔网逮住的玳瑁

被城市灯光迷惑的棱皮龟

▶ 第 26—27 页

海龟救援

2010 年，墨西哥湾发生了重大石油泄漏事故。石油裹住了海生动物，还被冲到了海滩上。石油对人类和野生动物都很危险。

被石油裹住的大西洋里德氏龟

路易斯安那州海滩上的石油

▶ 第 28—29 页

路易斯安那州负责海龟救援项目的人们挽救了很多海龟的生命。

救援人员把海龟清理干净，还给它们使用了药物。人们照料它们，直到它们可以回到大海。

大西洋里德氏龟

保护海龟

你不需要在海龟医院工作也能帮助海龟。你可以做下面这些事情来保护它们的安全。

捡起海滩上的垃圾。

不将气球放飞到空中。（它们最后往往落到海里。）

关掉能在海滩上看到的灯。

阅读海滩上的警示标语。避开海龟的筑巢区域。

与你的同学分享你了解到的有关海龟的知识。

▶ 第 32 页

词汇表

藻类植物： 生长在水里或水边、没有茎或叶的简单植物

濒危： 有灭绝的危险

幼体： 刚刚从卵里出来的、年幼的动物

本能： 动物天生就知道该怎么做的行为

捕食者： 杀死并吃掉其他动物的动物

爬行动物： 长着脊柱和鳞状表皮的卵生冷血动物